Plants in north-east Highlands

— timing of blaeberry growth, tree regeneration, land use, plant orientation

Adam Watson

Publication of this book was aided by the generous sponsorship of Bert McIntosh of Crathes near Banchory in Scotland and McIntosh Plant Hire (Aberdeen) Ltd, Birchmoss, Echt, Westhill, Aberdeenshire AB32 6XL, www.mphltd.co.uk

Published by Paragon Publishing, 4 North Street, Rothersthorpe, Northants, NN7 3JB, UK

First published 2014

Clachnaben, Crathes, Banchory, Aberdeenshire AB31 5JE, Scotland, UK
adamwatson@uwclub.net

ISBN 978-1-78222-189-0

Book design, layout and production management by Into Print
www.intoprint.net +44 (0) 1604 832149

Printed and bound in UK, USA and Australia by Lightning Source

Contents

Chapter
1 Timing of blaeberry growth on Scottish alpine land 5

2 Tree regeneration on moorland of Dee and Don 16

3 A critique of 'Norway and Scotland: a study in land use', a report May 1993 41

4 Highland sporting estates and land 'impoverishment' 52

5 Orientation in plants and animals 54

6 UK policies on Biodiversity and Sustainable Forestry 67

General introduction

The papers rest on observations I happened to make in the course of fieldwork on other projects, chiefly red grouse and ptarmigan. They were not studies with aims defined at the start. Nonetheless I made observations over many years with broadly similar annual effort. Results may therefore be of interest, for published data on these topics by other observers are scanty.

Chapter 1. Timing of blaeberry growth on Scottish alpine land

Summary

The timing of unfolding buds of blaeberry (Vaccinium myrtillus) was recorded at two hills on alpine land in north-east Scotland, in combination covering spring 1951–2013. On both hills it was correlated positively with mean air temperature in spring at Braemar in a nearby valley. There was a negative correlation with the snowiness of the previous winter as measured by the number of mornings when snow lay at Braemar and the number of downhill skier-days at Glenshee Ski Centre, a ski area that included the study areas on Cairnwell high (alpine) and low (sub-alpine). The longest run of data, on the two Cairnwell areas covered 1964–2013. During this run as a whole, the timing of growth was unrelated to the calendar year, and likewise during the run's first half in 1964–88. During the second half since 1989, by contrast, it was negatively correlated with the year, coming earlier as the years passed. This was a period of milder climate, shown by trends towards warmer air temperatures in winter and spring, fewer mornings with snow lying, fewer days of air frost and fewer skier-days. Timing varied more in 1989–2013 than in 1964–88. Five recent winters have been relatively cold and snowy, so a cooling period may have disrupted the trend to warmer temperatures. Timing was late in these years, and latest of all on Cairnwell alpine land in 2013, but came earliest of all during 2012 during an exceptionally warm March.

Introduction

One effect of a warmer climate is that plants start growing earlier in spring. This paper is a contribution to the subject, based on annual records over many decades. It documents data on the timing of blaeberry growth on alpine land in north-east Scotland during 1951–2013. Below, I test whether timing showed any trend over the years, whether it was correlated within years on different areas, and whether its variability changed.

Study areas

These were Derry Cairngorm in the Cairngorms massif north-west of Braemar village, and 20 km away the Cairnwell in the Mounth massif south of Braemar. The Cairnwell's upper three-quarters comprised alpine land, the lower quarter being sub-alpine or uppermost moorland. For brevity below, I call these Cairnwell high and Cairnwell low. Braemar lies by the river Dee, 90 km west-north-west of Aberdeen. Study areas have been described and illustrated in maps (Watson 1965, 1979; Watson et al. 1998). Cairnwell low stood at 650–760 m altitude, Cairnwell high at 760–930 m, Derry Cairngorm at 840–1150 m.

Methods

During winter, blaeberry on alpine land is a leafless dwarf woody shrub with greyish-green stalks that bear small buds. In spring, the buds swell and turn a lighter yellowish-green, shortly to become leafy new growth. I noted the date when buds showed the first unfolding of leaves, a conspicuous event when seen on close inspection.

Other species would be suitable. For instance, heather shows a pink spot at the tip of each newly growing shoot-tip, and crowberry a white spot in the centre of the uppermost whorl of leaves on the shoot-tip. I chose blaeberry, because on alpine land it is leafless in winter, so any new unfolding leaf is more conspicuous and hence less easily overlooked than on heather or crowberry. Growth in crowberry starts marginally earlier than in the other two species, up to a week but usually just a day or two.

I recorded timing as the number of weeks after 20 April, to the nearest week, occasionally to the nearest half-week if there were enough visits to justify this or when growth began within a day or two over a big span of altitude. I chose 20 April at the start of the study, before the more recent changes when springs tended to be milder and growth occasionally began before 20 April. In such cases of growth starting before 20 April, I used the number of weeks, but inserted a minus value.

Timing varied within an area, usually some days later at the top than at the foot, and it varied with aspect. Heavy browsing by domestic sheep, red deer (Cervus elaphus), mountain hare (Lepus timidus), ptarmigan (Lagopus mutus) and red grouse (Lagopus lagopus scoticus) delayed the onset, because browsed shoots had to form new tips and leaf-buds before growth could start. Browning during cold dry weather in winter and spring killed many shoots (Watson et al. 1966), delaying growth in affected plants.

Where soil under shallow snow had not frozen, heat from the ground melted the underside of the snow, creating an airspace, and buds there often showed the first signs of swelling, but did not reach the stage of the first leaf starting to unfold. Prolonged fresh snow and hard frost delayed growth. So did a greater depth of snow remaining in spring from a snowy winter, because warm air could not reach plants until the snow had gone.

The record was a mean for a study area. In a few years when very warm weather came to all altitudes simultaneously, I found timing to be the same at all altitudes (Watson 1965). More usually, a delay occurred with altitude, sometimes several weeks.

Mean monthly temperatures came from a climatological station at 339 m altitude (327 m since June 2005) at Braemar (Meteorological Office (MO) Monthly Weather Report and MO later data), as did the number of mornings with snow lying there (defined as snow covering at least half of the ground) from September to May, and the number of days with air frost from October to April. The MO has operated an automatic weather station on the Cairnwell summit at 933 m since 1993, but records are sometimes incomplete for various reasons. From the data I used March records for the small number of years with complete data for March.

Since 1962, Glenshee Chairlift Company has run a downhill ski-centre with car parks and the main base buildings at the foot of the Cairnwell at 645 m. Several pistes lie on the Cairnwell as well as other hills around the centre, served by ski-tows and chairlifts. In recent decades the Company has made artificial snow, which increased skier-numbers on snow-free days. However, the area involved with artificial snow-making covers a minute

Dying birches, lower Glen Ey, 11 December 1983 F

proportion of the ski area, and so the contribution of extra skiers on man-made snow to the cumulative totals of skier-days is not material. Since 1979 the Company has recorded the number of skier-days in each winter's ski-season from October to May (December to April inclusive are the only months in almost all seasons). Skier-days are like man-days, e.g. 100 skiers on one day and 50 on the next two days amount to 200 skier-days. The total number per given ski-season is a useful proxy index on the snowiness of the winter in terms of snow-lie and its duration.

Results

First dates in relation to calendar year

Fig. 1 shows the runs of data, and Table 1 the mean dates and other statistics. In the 1964–2013 runs at the Cairnwell, correlation coefficients in comparisons with the calendar year were weak and far from significant (Table 2).

When data were split into each half of the run, correlations remained insignificant in the first half (Table 2). However, they shifted between the first and second halves, towards larger values and negative signs, significant in the second half. This shift towards earlier dates in the two most recent decades fitted the expectation that plants would start growing earlier during recent years of milder climate.

During the second half, there were also trends in air temperature and snow-lie (Table 2). These involved warmer temperatures in April and fewer winter mornings with snow lying at Braemar (though neither significant), and fewer skier-days at Glenshee Ski Centre. It can be seen in Fig. 1 that the timing in 2008 and 2010 was quite late, following two snowy winters, and the winters of 2009 and 2011 were also snowy. Hence a cooling period may have begun, disrupting the warming. Against this is the fact that the timing in 2012 was the earliest of all, but the timing in 2013 came the latest of all years on alpine land, with no growth until June.

Variability of first dates amongst years

Variances for blaeberry growth during the second half exceeded those in the first half (Table 1). A commonly used index of dispersion (the variance divided by the mean) confirmed this. The test for the equality of variances was significant at the Cairnwell areas (Table 3), thus pointing to a greater variability of timing in recent years.

Timing of growth compared with temperature and snow

Timing was negatively correlated with air temperature at Braemar, for Cairnwell high and low involving April temperature, and for Derry Cairngorm the May temperature (Table 4). Later timing at Derry Cairngorm was expected, because most of the study area lay at higher altitudes than the Cairnwell high area.

Timing at Derry Cairngorm was positively correlated with the number of winter mornings with snow lying at Braemar village (Table 4). Thus, for example, blaeberry on the hill tended to start growth later in springs that followed winters with many snow-mornings at the village.

At the Cairnwell areas, timing was

again positively correlated with winter snow-mornings at Braemar. There were strong correlations between timing and the number of skier-days in winter at Glenshee Ski Centre. These were expected, because the Cairnwell study areas lay within the downhill ski area, close to the main buildings and car parks, and popular downhill pistes traversed both the Cairnwell high and the Cairnwell low area.

Timing of growth compared on different areas

The question here was whether the timing of growth was correlated within years on different areas, e.g. whether all areas tend to be early, or late, in a given year. Comparisons of Derry Cairngorm and the Cairnwell showed almost no association during the short run of common years (Table 5). Most of these years had snowy springs, when Derry Cairngorm carried far more snow than the Cairnwell, and so a correlation of timing on the two hills was not expected. However, timing on the Cairnwell high and the Cairnwell low, covering a much longer run of years, was strongly correlated, positively. A stronger relationship is to be expected where two areas adjoin, as at the Cairnwell, than where they are 20 km apart in different massifs as at Derry Cairngorm and the Cairnwell.

Discussion

Possible factors confounding the expected impacts of warmer climate

Although the timing of growth was positively correlated with April temperature at Braemar during the period of warmer climate since 1989, this accounted for only a minor proportion of the variance in timing. Other factors must be involved.

One that might reduce the impact of warmer climate is less snow-lie. Less snow-lie had already been reported for several Scottish settlements including Braemar (Harrison 1996), and since the late 1980s has affected all five Scottish downhill ski centres, with Glenshee Chairlift Company going bankrupt in 2004 (Watson et al. 2005). Snow acts as an insulating blanket and in cold weather keeps the surface soil horizons warmer than they would be if no snow lay. Hence less snow-lie might result in cooler soil temperatures in the surface soil horizons where most plant roots occur on alpine land, and thus a greater delay in plant growth than expected from mean air temperature alone.

Furthermore, less snow exposes blaeberry to wind and consequent damage from winter browning, which delays the onset of growth. Also, less snow results in blaeberry being exposed to heavier browsing, and browsing is known to delay new growth. Densities of red deer and mountain hares on the Cairnwell have risen substantially during the study, and their impacts have been marked in the last decade (A. Watson, unpublished data). In addition, less snow exposes blaeberry differentially as against heather, because blaeberry abounds most in snow-patch hollows (Rodwell 1992).

To sum up, the positive correlation between timing at the Cairnwell and temperature at Braemar during the warmer climate since 1989 accounted for only a minor proportion of the variation. The large unexplained proportion may be due to other factors associated with a warmer climate. These may be counteracting the expected simple effects of plant growth starting earlier. Yet other factors may confound the interpretation, especially the rise in densities of red deer and mountain hares, which themselves may also be associated with milder winters and less winter mortality.

This exemplifies a general problem with research on a topic such as climatic change, where the change is an experiment without a control. Associated factors that have not been the main object of a given study, and may not even have occurred at the start of the work, may nonetheless affect or confound the results.

Timing in relation to warmer climate
Because of climatic warming in the last two decades, an earlier response by plants might be expected in 1989–2013. The results above reveal an earlier response in the timing of blaeberry growth, and a tendency for greater variability than in 1964–88.

Acknowledgements

I thank Stuart Rae for some supplementary observations in three years, and managers of Glenshee Ski Centre and Marian Austin the manager of Nevis Range Ski Centre for data on skier-days. Credit is due to the Meteorological Office for affording me unpublished data on snow-lie in recent years, via the BADC scheme.

References

Harrison, J. (1996). Changes in the Scottish climate. Botanical Journal of Scotland 49, 287–300.

Rodwell, J..S. (Ed.) (1992). British plant communities. Vol. 3. Grasslands and montane communities. Cambridge University Press.

Watson, A. (1965). A population study of ptarmigan (Lagopus mutus) in Scotland. Journal of Animal Ecology 34, 135–172.

Watson, A. (1979). Bird and mammal numbers in relation to human impact at ski lifts on Scottish hills. Journal of Applied Ecology 16, 753–764.

Watson, A., Miller, G.R. & Green, F.H.W. (1966).Winter browning in heather (Calluna vulgaris) and other moorland plants. Transactions of the Botanical Society of Edinburgh 40, 195–203.

Watson, A., Pottie, J. & Duncan, D. (2005). One Scottish snow patch survives until winter 2004/05. Weather 60, 165–166.

Watson, A., Moss, R. & Rae, S. (1998). Population dynamics of Scottish rock ptarmigan cycles. Ecology 79, 1174–1192.

No young pines due to deer browsing in old wood, Luibeg, June 1957

Table 1. Arithmetic mean, variance, index of dispersion (variance divided by mean), and minimum and maximum values in the timing of blaeberry growth (weeks after 20 April).

Study area	Years	n	Mean	Variance	Index	Min	Max
Derry Cairngorm	1951–72	21	4.31	1.46	0.34	2.5	6.5
Cairnwell high	1964–88	25	2.34	1.66	0.71	1.0	5.5
	1989–13	25	1.68	5.08	2.76	–3.5	6.5
Cairnwell low	1964–88	25	1.42	0.60	0.42	0.0	3.0
	1989–13	25	1.10	3.22	2.93	–3.5	4.5

Table 2. Timing of blaeberry growth, mean daily air temperature at Braemar, number of winter mornings with snow lying at Braemar, number of days with air frost at Braemar, and number of skier-days in winter at Glenshee Ski Centre, compared with the calendar year.

During the 1964–2013 run as a whole, associations with the calendar year were negative at the Cairnwell high and low areas (n = 50, respectively r = –0.210 and –0.259, P = 0.14 and 0.07).

	Location	Years	r	P
Blaeberry growth	Derry Cairngorm	1951–72	0.06	0.8
	Cairnwell high	1964–88	0.19	0.3
		1989–13	–0.655	0.0005
	Cairnwell low	1964–88	0.11	0.6
		1989–13	–0.670	0.0003
May temperature	Braemar	1951–72	–0.23	0.2
April temperature		1964–88	0.12	0.6
		1989–13	0.21	0.3
Winter snow mornings		1951–72	–0.13	0.6
		1964–87	0.33	0.11
		1989–12	–0.20	0.3
Winter skier-days	Glenshee	1979–13	–0.381	0.02

Table 3. Differences in the mean timing of blaeberry growth in 1964–88 and 1989–2013.

Study area	T	P	F*	P*
Cairnwell high	0.50	0.34	3.66	0.0041
Cairnwell low	0.32	0.44	5.93	<0.0001

*F test for equality of variances, $P < 0.05$ gives statistically significant support for assuming that variances between the two sets of years are unequal.

Table 4. Timing of blaeberry growth in relation to mean monthly air temperature at Braemar, number of winter mornings with snow lying at Braemar*, and number of skier-days in winter at Glenshee Ski Centre.

	Study area	Years	r	P
May temperature	Derry Cairngorm	1951–72	–0.476	0.029
April temperature	Cairnwell high	1964–13	–0.479	0.0004
	Cairnwell low	1964–13	–0.437	0.0015
Winter snow mornings	Derry Cairngorm	1951–72	0.474	0.030
	Cairnwell high	1964–12	0.325	0.023
	Cairnwell low		0.306	0.033
Winter skier-days	Cairnwell high	1979–13	0.453	0.0063
	Cairnwell low		0.469	0.0045

*No Braemar data from summer 2005 onwards, data at nearby Balmoral used. Timing on Cairnwell high and low areas was correlated with mean March temperature on Cairnwell summit ($n = 14$ years with complete data, r, $= -0.656$ and -0.708, $P = 0.0108$ and 0.0046).

Table 5. Timing of blaeberry growth compared on different study areas.

Study area	Years	r	P
Derry Cairngorm and Cairnwell high	1964–72	–0.140	0.72
Derry Cairngorm and Cairnwell low		–0.006	0.99
Cairnwell high and low	1964–13	0.894	<0.0001

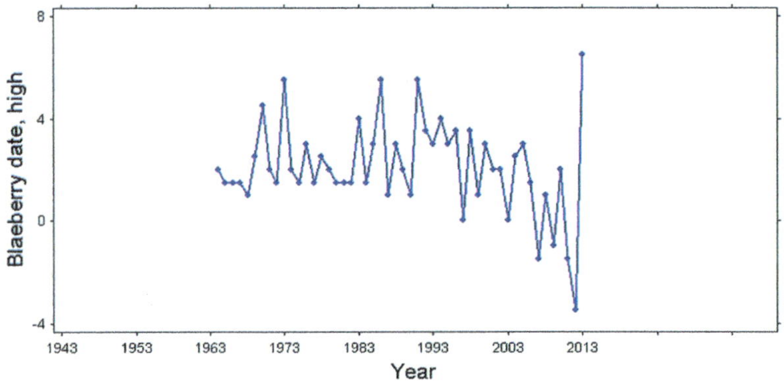

Fig. 1. Timing of blaeberry growth (measured in no of weeks after 20 April) on Cairnwell high, the study area for ptarmigan at the Cairnwell, 1964–2013.

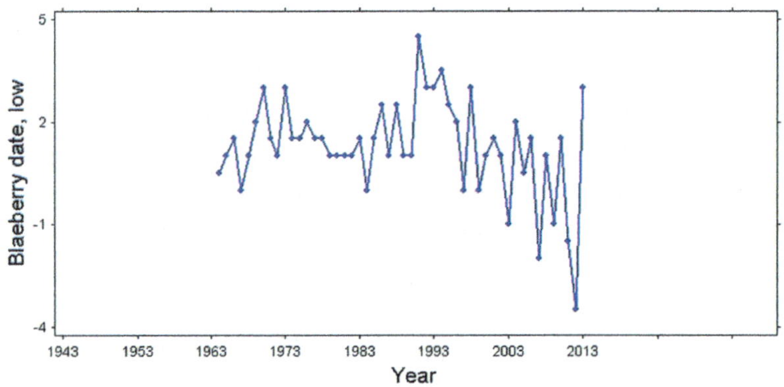

Fig. 2. Timing of blaeberry growth (measured in no of weeks after 20 April) on Cairnwell low, the study area for red grouse at the Cairnwell, 1964–2013.

Chapter 2. Tree regeneration on moorland of Dee and Don

Summary

In summer 1985 I did a total survey of natural tree regeneration on open upland, mostly heather moorland, in the Dee and Don river-catchments of north-east Scotland. The survey revealed large areas of regenerating Scots pine and birch, with considerable areas of rowan, willow and other species. Much of it was still scattered as young trees on moorland, but substantial areas had changed from moorland to woodland, mostly with young and middle-aged trees. In Deeside, 584 grid kilometre squares on the Ordnance Survey (OS) map contained natural regeneration on open upland, and 440 in Donside. Two kinds of maps are presented. One shows the approximate proportion of each such kilometre square that contained regeneration by a given species. The second kind shows for each such square that proportion of the area containing regeneration that was occupied by a given species. Most of the area with regeneration was on grouse moors. There was substantial regeneration on lowland peat mosses. Deer forests with high numbers of red deer (Cervus elaphus) had virtually no regeneration, as had large areas of rotationally burned grouse moor and smaller areas of heavily browsed moorland near farms with large stocks of sheep and cattle. On some estates, managers since 1985 have reduced or removed much of the regeneration of seedling trees by pulling them out, of young trees by burning, and of older trees by felling them and leaving them on the ground. The implications of this natural transformation of moorland to boreal woodland are discussed. There are obvious implications for land use, nature conservation, wildlife, scenery and forestry policy, and other factors such as water catchment.

Background

Most of the lower moorland in Britain and Ireland was originally forest (Tansley 1953). Forest clearance occurred, largely due to man, centuries ago (Bunce & Jeffers 1977). Since then, the ground has been kept open by burning and by the heavy browsing of unnaturally dense stocks of sheep, cattle and red deer (Cervus elaphus). It follows that when pulling, burning and heavy browsing of tree seedlings cease or become greatly reduced, the lower moorlands should revert to forest, particularly near existing trees that can supply seed sources. The present study documents a notable case of such reversion towards forest over a large area in north-east Scotland.

In summer 1985, I carried out a survey of natural regeneration of trees on open upland in the entire Dee and Don catchments in Aberdeenshire and north Kincardineshire. This paper describes the survey, presents the main data, and discusses implications for land use, nature conservation, wildlife, scenery, and forestry policy.

Study areas

The fieldwork did not rest on a sample but involved a total survey of tree regeneration on open unfenced upland in the Dee and Don catchments of north-east Scotland. I did all the fieldwork, so the same observer was involved throughout. The potential upper limit of woodland was judged by actual limits of old woodland in some places, and by the growth form of trees in others. It varied from 640 m in the Braemar area to 400 m in the more exposed and more oceanic parts of lower Deeside and Donside. Occasional trees occur above these altitudes, including high alpine ground in the Cairngorms, but such trees are rare, do not persist, and are of negligible importance in considering the habitat types that are present.

I classified land in the two catchment areas of Dee and Don into four types according to its suitability or otherwise for natural tree regeneration on open upland. Land type A was that above the potential upper woodland limit, and contained no woodland. Type B had no open upland, and consisted of agricultural fields, settlements, tree plantations, surfaces of lakes and rivers, etc. It was not surveyed.

Land types C and D were open uplands with potential for tree regeneration. Type C was virtually without regeneration, because regeneration was prevented by high numbers of browsing sheep, cattle and red deer and/or burning. Where type C ground a) had been deer-fenced and large browsing mammals excluded, or b) involved cliffs, gorges, and rocky stream banks, tree regeneration was common. I visited type C thoroughly, but it does not feature in this Chapter because virtually no regeneration occurred there, except in cases a) and b) as mentioned in the previous sentence.

Land type D was open upland where appreciable tree regeneration was found, and most of this report is about type D. Type D included unfenced, usually semi-open, native woodlands next to open upland, if there was natural regeneration there. It covered the first fenced ground adjoining large continuous tracts of open upland, apart from tree plantations, if the fence had been erected in the previous ten years. This was useful in showing cases where regeneration had previously been nil but had occurred since fencing. Type D included isolated open areas of uncultivated ground, usually hilltops projecting above fenced fields or woods. Type D also covered isolated lowland peat mosses on valley bottoms, which were fenced off from nearby fields or woods. These cases were included as the peat mosses are of considerable conservation interest. They were useful in showing ground where regeneration had recently (in 1985) been more vigorous than in previous decades, probably due to less browsing by sheep and cattle and less burning.

Areas that were fenced off and had been densely planted with trees by the time of the survey in summer 1985 were included in type D if they were known to have been open unfenced upland since 1980, and also contained naturally regenerated trees which had survived the ground preparation for afforestation; such trees were larger than the planted ones and different in species from them. The reason for including such areas was that they showed where trees had been regenerating but where further spread of this had now (1985) stopped. Areas of former woodland which had long been felled and become open upland were also included in type D. I included four cases where trees were regenerating naturally on enclosed agricultural fields, because this was of interest in being so unusual.

Tree plantations were excluded, as were natural or semi -natural woodland in the midst of heavily enclosed low ground (except for peat mosses and former open upland areas as mentioned above). Also I excluded narrow forest rides, and wider strips cleared through forest for pipelines and electricity lines. Strips cleared for pipelines and electricity lines usually show rapid tree regeneration, though often at an earlier, more pioneering stage of the succession, such as mainly birch seedlings in a wood with much pine and birch (for Latin names of tree species, see Table 1).

I included areas with relic trees from former plantations, after checking with local inhabitants. Such cases were usually obvious, involving European larch, Norway spruce and Scots pine at high altitudes, often along the upper fringes of a former wood showing cut stumps. Former plantations occasionally involved hardwoods such as oak and beech scattered in lines or as individuals rather than in larger patches as seen in natural oak woods.

The vast bulk of the area in land type D comprised open unfenced upland. The other sub-categories within D, mentioned above, made little difference to the total. I included them because they demonstrated points of interest. The scope of the present report was limited, but data on these sub-categories could be extracted in a more detailed investigation in future. The full data base has been put on a computer and archived on discs.

Methods

Meaning of 'regeneration'

Natural tree regeneration on open upland in the area was obvious. Most of it involved a spread of young and to a lesser extent of middle-aged trees over the last few decades on to ground which I knew to have been treeless previously and which I knew had not been planted. 'Known' rested on my experience of the area since 1943, aided by discussions with local people who had lived there back to the end of the 1800s, and information on old OS and other maps.

Another feature of the natural regeneration was that trees even within a small area such as a hectare were of highly uneven density and to a lesser extent age. The only snag arose over trees more than 80–100 years old, which were scattered and uneven in density because of past felling. Such trees often looked natural at first sight, but might have been planted; in most cases it was known that they had been. In some uncertain cases, the regular distribution of stumps frequently showed that they had been planted.

Only in a few cases with very old Caledonian-type pinewoods was there any doubt. To remove this doubt completely would require detailed study of past estate records and maps, which would be a major historical research study in its own right, and even then would not succeed in the many places where no such old records and maps are available. However, such cases were negligible (far less than 1%) in terms of the total area of ground with regeneration. In cases where old trees were found to be remains of a former plantation on unfenced open upland, I included

Pine tree on rock out of reach of deer, Carn Crom, November 1985

these in the survey if they grew along with naturally regenerated younger trees.

Places visited during the survey but not quantified by the survey included old former plantations which once had fences or dykes that have since broken down, and where there had been no regeneration of younger trees in recent years up to 1985. These all lay on land type C, as did all unfenced natural old woods with virtually no regeneration and next to open upland.

Throughout the study, all areas of open upland or that had recently been open upland were visited by the same observer to locate and establish the characteristics of tree regeneration. I plotted the location of regeneration on OS maps of 1:10000 scale in the field. Areas of regeneration for each tree species, for different tree density, and for tree age, were marked on maps in the field.

Many km squares wholly comprised land type A (see Study Area) above the potential woodland line. Others were wholly of land type B, occupied by arable fields, plantations, villages, lakes etc. Many squares consisted entirely of land type C, on open upland where virtually no natural tree regeneration occurred, but where it could occur if constraints by burning and browsing were reduced or removed. The remaining land type D, where I found natural tree regeneration, was the land type that predominated in this study.

It will be possible in future to list all squares consisting wholly of one of the three land types A, B, and C. If such data for land type C were to be used along with the tree regeneration maps, they would allow one to find those main areas where natural tree regeneration on open upland could not occur, and those where it does not occur but where it potentially could occur. In many cases a given square was not wholly of land type A, or B, or C, but mostly occupied by one type and partly by another. The scope of this report did not allow for data abstraction and presentation for such areas without regeneration. However, it would be possible in future to produce a more complete regeneration analysis in relation to land use in all squares.

Where two or more species were regenerating in association, I recorded the species in order of their preponderance. Thus, if regeneration on an area was mainly of pine with some birch and even fewer willow trees, what I plotted on the field map showed them in that order.

Density was scored as 0, virtually no trees; 1, scattered trees; 2, trees in clumps; and 3, woodland. Score 0 did not signify no trees, but did signify no large trees. Areas with score 0 might well have occasional small seedlings in the ground vegetation, but no saplings or larger trees of man height or higher, except on cliffs, gorges, rocky stream banks and inside fences. In score 3, the trees were sufficiently dense to obscure all or nearly all of the heather or other low vegetation below them. In score 1 the area was obviously still mostly open, and at a distance of 1 km often appeared wholly so unless binoculars were used. The characteristics of score 2, for medium tree density, were intermediate (i.e. neither what most people would regard as an open moor nor a wood), and hence harder to define, and only the observer could judge, based on experience gained through the study. I changed a few of the

Self-sown Scots pines on heather moorland, Glen Cat, October 1985

Self-sown Scots pine and Norway spruce, Corse Hill, February 1986

earlier scorings in the light of this greater experience.

The survey was done when all trees were in leaf. In winter, when broad-leaved trees and larches are leafless, more of the moorland vegetation behind them can be seen, as compared with pines of the same size and density. Hence the density of broad-leaved trees and larches would seem lower in winter than in summer.

I estimated the age of coniferous trees by noting the annual increments of vertical growth and annual whorls of side branches, but this method did not apply to broad-leaved species. However, the age class could be assessed from the size and growth form. Size can be misleading on its own, as an old tree at high altitudes, on a dry rock, or in thick wet peat can be as short as a young tree, but it has a different growth form. My experience of the growth form of broad-leaved trees of known age, and my age classification of coniferous trees growing nearby, were useful in deciding age categories of broad-leaved species. The term 'age' below is used for convenience to cover the above points.

I scored age as 1, young; 2, middle-aged, covering saplings and small trees intermediate in growth form; and 3, old, covering trees usually taller than 10 m. Score 3 covered a variety of types, such as with a rounded upper surface (Scots pine), or very tall (larch and spruce), or with most branches dead (some willow and other species).

Trees in bogs and at high altitudes had small height and stature, even though known from local informants to be the remains of old plantations. Such trees were easily distinguished from young trees, however, and have been included in the 'middle-aged' category. Another useful rule of thumb, which could be checked in numerous cases by observing the effects of moor burning, was that trees in age-score 1 would all burn in a moorland fire; pines, spruces and larches would be destroyed, and birches and some other hardwoods would often appear again later, but springing from the root. Many pines of age-score 2 die after moor fires, but some of the larger ones survive. Very few old pines die after moor fires. Obviously these remarks do not apply to woodland fires, which have more serious impacts.

The effects of damage by cattle, sheep, deer and other browsers provided another useful rule of thumb. Trees of age-score 1 could easily be destroyed above ground level. This caused complete destruction of a conifer and destruction after repeated browsing of hardwoods. Many age-score 2 trees survived browsing, though bark stripping or rubbing could kill them and also kill occasional trees of age-score 3.

In many cases a combination of age classes occurred on the same ground. For example, a stand of trees might be predominantly old, but some young trees and even fewer middle-aged trees might occur in the same stand. The record would show old, young, and middle-aged, in that order.

Isolated individual trees on open upland in sites far from other trees were also noted. I plotted their locations on maps, and recorded their species and age. Usually they were of middle age, occasionally old. I considered them important as seed sources for potential regeneration after possible future reductions of browsing animals.

Activities likely to be detrimental

Self-sown Scots pines, Crannach, July 1985

to natural tree regeneration were noted during the survey. This frequently involved heather burning, sometimes on a large scale and concentrated upon areas with many young trees. In some places, larger trees were chopped down and left on the ground. Alteration of heather moorland to agricultural grassland had occurred in some places, destroying young trees but leaving some older ones. Intensive, large-scale afforestation was an important cause of loss, and a lesser one was underplanting of birch or other woods with coniferous trees.

In some places, remains of earlier regeneration remained, but with no recent regeneration, indicating that conditions had changed. In this last case, the only factor common to all the examples was obviously increased numbers of sheep, cattle or red deer, but the scope of this report does not allow for inclusion of stocks of such browsing animals.

If I saw that the density of trees or the species composition changed during the course of the survey, this always followed one of the above changes in land management. If so, I revisited these areas and plotted new details of afforestation etc on maps in the field, so that the survey was complete at the end of 1985. Changes since then have been noted on maps and could be produced eventually as a supplement, but are not in this report.

Some areas had been changed since the 1:10000 OS maps were published in 1974, by turning moorland into agricultural grassland, and most frequently by afforestation. Ground preparation for afforestation had sometimes destroyed all former tree cover, and I noted this. In most cases, this was known from prior experience of the area, often backed up by photographs. In a few cases I assumed it from the existence of old stumps. In other cases, afforestation had left some natural trees, still obvious by their greater age. In other cases, areas had only recently been ploughed for afforestation or were being ploughed during the survey period, although no trees had yet been planted. In all such cases, I recorded this as afforestation, and all such areas were planted within the following six months.

The OS map symbols were not reliable as a guide to tree cover in 1974 or since. Some areas shown as woodland on OS maps had no trees or virtually none in 1985, and the evidence in the survey was that they had been age-score 1 trees which had been destroyed since by moor burning. Areas shown as woodland on OS maps were often given an age-score of 3 by me in the field, sometimes only 2. Map symbols for conifers and for hardwoods were also unreliable. In some areas that had conifers only, the OS map showed hardwood symbols only, and vice versa.

I drew estate boundaries on maps. Most had been found in the course of another study of bulldozed hill tracks (Watson 1984). In some cases I asked local informants, and in a few remaining cases decided the boundaries from Callander's (1987) study of landownership and his unpublished notes (R. Callander, personal communication). As land changes hands over the years and some estates sell some areas, the landownership pattern is not static. It can be taken as reasonably accurate in 1981–82 when the survey of tracks was made, and it still applies over most of the area, particularly on the large estates, which have mostly remained unchanged.

The main land uses on hill ground owned by these estates was also known from the survey of hill tracks. As a result, it would be possible to use the data on tree regeneration to show which estates had most natural regeneration of trees on open upland, which kinds of main land use on open upland were associated with different amounts of natural regeneration, and which estates were reducing or removing trees on a large scale.

Such comments should not be taken necessarily as criticisms of land management on particular estates, but are simply objective statements of what is there and what is happening. If landowners, public or private, wish natural trees they leave the trees, and if they wish an area of developing trees on grouse moor to be open moorland then they remove the trees. Examples of both were found on several private estates, and also on land under public management, for instance on the National Nature Reserve at the Muir of Dinnet, where trees have been left on most parts of the Reserve, but have been cleared by cutting and burning on other parts so as to maintain moorland vegetation of conservation interest. The present report does not present these data in relation to different landownership patterns and different land use, but this will be given in future.

C. Sampson calculated the area of each individual patch of regeneration drawn on the field maps to the nearest quarter hectare, using a grid overlay. She then gave each isolated tree on open upland an allocated ' area ' of 0.001 hectares. Within each kilometre square separately, the areas of trees for each category of species, density, and existence of detrimental activities, were transcribed on to paper and coded, ready for building a computer database. I added an additional variable, the name of the estate or landowner on whose ground the regeneration had occurred, and she separated the data within each square according to landownership.

The tree regeneration database covers only those kilometre squares that had regeneration on open upland or on ground that was until recently open upland, as described above. Consequently a complete coverage of the 100 hectares of the kilometre square occurred only if the whole square was under regeneration, which was not usually the case. Hence, interpretation of the tree maps must take into account the area of each species within the area containing regeneration in that kilometre square. For example, 1 hectare of regenerating pine in a kilometre square with a total regeneration area of 1 hectare would appear to be far more important than 1 hectare of pine in a kilometre square with a total regeneration area of 100 hectares, 99 of which consisted of other regenerating species.

TM. Hinge divided the completed database into a subset database that contained all kilometre squares with regeneration in Deeside, and a database similarly for Donside. As some squares straddling the watershed of Deeside and Donside had records for both water catchments, the squares appear in both databases.

M. Hinge used a computer with both databases to sort kilometre squares into ascending order. After a list of all species recorded had been compiled (Table 1), it was simple to establish the presence or absence of each species in all the kilometre squares that had regeneration.

The calculation of area occupied by each species in each kilometre square was simple where the species occurred in pure stands. However, two or more species often occurred in association. Because I recorded no information on the amount of each species in the association, other than their order of preponderance, M. Hinge and I weighted the area occupied by each species according to its position in the order and also the number of species in the association (Table 2).

To establish which km squares had new, or sustained, or old regeneration, I defined each such class. New regeneration was easy to define, as those stands with only young regenerating trees. In some old stands where trees were gradually dying and not being replaced by young trees, middle-aged trees often grew, or a combination of middle-aged and old. Sustained regeneration occurred in areas that had all age classes, or else combinations of young and middle-aged trees along with old trees.

Because the survey covered the total population of natural tree regeneration on open upland in the two catchments, statistical analysis based on samples was unnecessary and inappropriate.

Results

General findings

The total area of regenerating trees on open upland in Deeside amounted to 16413 hectares in 584 kilometre grid squares, and in Donside 9671 hectares in 440 kilometre grid squares. The total areas with single-species tree regeneration in Deeside and Donside are shown in Table 3.

M. Hinge set the computer to map km squares with regenerating trees. Total

Self-sown Scots pines on Benaquhallie, showing browse line by cattle, March 1979

Grouse butt in self-sown Scots pines at Glen Cat, September 1985

shading within each pie represents one hundred, irrespective of whether the measure was percent or hectares. The main tree species for which maps have been drawn are Scots pine, birch, rowan, willow, larch and Norway spruce. The distribution of the less common species is described without mapping: alder, ash, aspen, beech, hawthorn, holly, oak, dog rose and yew.

For each species mapped, M. Hinge drew two maps. The first map, ' Percent of 1 km square', shows the approximate area in hectares occupied by the species, as indicated by the amount of shading within each pie for that particular kilometre square. The second map, marked ' Percent of the area with regeneration in 1 km square', shows the proportion of regeneration for a given species in the total area of regeneration within that kilometre square.

The survey revealed large areas of good natural regeneration of trees, mainly of Scots pine and birch, but with a considerable variety of other species in small amounts. The vast bulk of the area with regeneration occurred on grouse moor, in both Deeside and Donside separately. Some small areas with regeneration lay on agricultural upland that was not grouse moor, in places where heather moorland had been reclaimed into agricultural grassland without destroying middle-aged or old trees that had been there before the reclamation.

The largest areas of regenerating trees on open upland in Deeside were on grouse moors north of the River Dee between Crathie and Aboyne, including lower Glen Gairn, Pressendye, Craiglich and Mortlich, and the Hill of Fare, and south of the river between Glen Tanar and Kerloch. Tilquhillie near Banchory had much regeneration in the early 1980s, but all this ground had been intensively afforested by the time of the survey in 1985.

Compared with Deeside, Donside had smaller areas with regenerating trees on open upland, but substantial amounts at Culquoich, Glen Kindie, Littlewood, Coiliochbhar Hill, Bennachie, Craiglich, Corse Hill and Corrennie. Knockespoch on the Correen Hills and the west side of Glen Buchat had much regeneration in the early 1980s, but most of this ground had been afforested by the time of the survey.

Compared with Donside, Deeside had a larger number of kilometre grid squares containing regenerating Scots pine, birch, rowan, and willow, but in Donside a larger number had larch, Norway spruce and beech than in Deeside (Table 1). Open upland in Donside also had a larger area where the regeneration comprised pure rowan, pure larch or pure beech (Table 3). The tendency for more rowans in Donside than in Deeside may reflect the generally better soils in Donside.

Distribution of lesser tree species on open upland in Deeside

Alder: OS grid kilometre squares NO 3697, 4399, 4698, 4899, 5791, 7499, 7698 and 7798; and NJ 4401, 7810 and 7909. The largest area of pure alder (1 ha) occurred in NO 7798. No areas of aspen in association occupied more than 1 ha.

Ash: I saw only one area of ash (N0 7193), at low density in a middle-age birch/rowarl/ash association (0.75 ha).

Aspen: NO 3497, 3799, 4499, 4799

and 5191; and NJ 3300 and 4300. The largest area of pure aspen (0.75 ha) occurred in NO 4499. All other areas in association were 1 ha or less. I found no aspen areas of low density.

Beech: N0 6399 and 8897; and NJ 4702, 4807, 5606, 5705, 5803, 5804, 5907, 6007, 7208, 7307 and 7308. The largest area of pure beech occurred in NJ 5804 (4.75 ha), of old trees in low density. All areas of beech had old trees.

Hawthorn: Two areas of low density and young age occurred in N0 4998 and 4999, both in the association rowan/birch/pine/hawthorn, and occupying 8 and 6.75 hectares respectively.

Holly: N0 5090, 5190, 5590, 5591, 5690, 5791, 6288, 6289, 6294, 6388, 6389, 6394, 6397, 6489 and 7189; and NJ 6003. Thc four largest areas of pure holly (24.75, 15, 11.5 and 9 ha) occurred in N0 6389, 6288, 6489 and 6388 respectively. In all sites where holly occurred, it grew either in a pure form or was the least abundant in an association of species. I saw no dense holly stands, and no old holly trees. Most were young stands or stands of young trees with a few middle-aged trees.

Oak: N0 3497, 3696 and 5690; and NJ 4901, 5804 and 5907. The largest area of pure oak regeneration (1.5 ha) occurred in NJ 5804, with old trees at low density. In N0 3696, dense old trees occurred in an oak/pine/birch/larch association (22.3 ha).

Rose: N0 3696, 5590, 5591, 5690 and 7189. The only area of pure rose occurred in N0 5590 (0.25 ha), and the largest association (rowan/willow/birch/rose/pine) in N0 5690 (6 ha).

Yew: No regenerating yew was found.

Distribution of lesser tree species on open upland in Donside

Alder: Two alder patches (NJ 8108 and 8510) had old trees occupying small areas (0.25 and 0.125 ha respectively).

Ash: NJ 5217, 5530 and 5630. Ash on all areas, whether pure or in association, grew at high density. The biggest area of ash in association (birch/rowan/ash) occurred in NJ 5630 (6 ha).

Aspen: Only one area of aspen was found (NJ 4818), in an association of low-density, old-aged pine/aspen/oak (0.75 ha).

Beech: NJ 4315, 4415, 4918, 5116, 5216, 5217, 5218, 5316, 5317, 5318, 5417, 5418, 5423, 5514, 5530, 5607, 5630, 5631, 5713, 5731, 5806, 5813, 5907, 5911, 6535, 7413, 7913, 8023, 8024, 8108, 8124, 8409 and 8415. The largest area of pure beech occurred in NJ 5216 (3 ha), whereas in NJ 4415 the largest association (8 ha) comprised pine/oak/beech. Most of the beech areas were of middle or old age. Only NJ 5316, 5216, 5318 and 5630 had some young regeneration.

Hawthorn: No regenerating hawthorn, holly, or rose was found.

Oak: NJ 4315, 4818, 4918, 5217, 5218 and 5530. The largest pure stand of oak (2.5 ha) occurred in NJ 5530, whereas the largest area of oak (13 ha) in association (pine/larch/oak) was in NJ 4315. Virtually all areas with oak had middle to old-age trees. Square NJ 4918 had six areas (total 32.5 ha) of pure oak and of oak in a pine/birch association.

Yew: One area (NJ 5320) of dense, old yew was found in association with conifers (1 ha).

Discussion

Areas with regeneration

Watson (1977) noticed that small Scots pines had colonised moorland on a large scale above the pine woodland at Geallaig Hill near Ballater, and that pines were regenerating vigorously near Dorback on Speyside and on Deeside at Glen Gairn, Crathie-Ballater, Glen Tanar and Kerloch, with birch spreading on the Muir of Dinnet (Nethersole-Thompson & Watson 1981). Brown & Wightman (1988) found that more than 50% of birchwoods established since 1947–49 on Deeside and still existing in 1983–85 had resulted from colonisation of moorland and rough pasture. My survey also found large-scale regeneration of Scots pine, and to a lesser extent birch, on open upland in Donside. Donside was also notable in having patches with good regeneration of rowan, and considerable regeneration of beech, larch and Norway spruce in places.

Data in the survey showed large areas of natural tree regeneration, which has changed many tracts of open upland to developing young or middle-aged boreal woodland. Over big areas, however, the regeneration comprised an early stage with the seedlings still highly vulnerable to fire. During and since the survey, staff on several estates were reducing or destroying much of this tree regeneration by pulling out seedlings, burning and felling. On some estates it was being removed by intensive afforestation, whereas other estates have not destroyed or reduced it.

The results of the survey show a major transformation of land over large areas. This is relevant to those interested in land use, nature conservation, wildlife, soils, water catchment, flooding, river fisheries,

Gamekeepers burned Scots pines, Glenfenzie, June 1976

Gamekeepers felled Scots pines at the Strone, Glen Gairn, September 1986

snow clearance on winter roads, scenic views, field sports, and walking and other recreation on open upland.

As well as large areas of open upland becoming transformed to natural woodland, the survey showed that many small lowland peat mosses have become less open and are developing towards woodland. In time this will tend to lower the water table and cause changes in the pool flora and fauna. I found four cases with trees regenerating vigorously inside fenced fields. These were so small in extent as to be negligible in terms of the total land area covered in the survey. Nevertheless they are of interest in showing that pasture fields with low usage by sheep and cattle have a potential to develop towards natural woodland, without any need for planting. This is not surprising in view of the well known colonisation of abandoned fields in North America by naturally regenerating scrub and trees (summary in Krebs 1978), but the point seems not generally appreciated in Britain.

Areas without regenerating trees

All old Caledonian pine forest relicts (Steven & Carlisle 1959) in the survey area showed virtually no regeneration outwith deer-fenced enclosures, except for Glen Tanar. The only saplings seen outwith fenced enclosures were growing on the upturned roots of wind-blown trees and on rocky gorges or cliffs, out of reach of red deer. Such regeneration was negligible, and inadequate to replace the forest in the long-term. The lack of regeneration this century has been attributed to many red deer. Gordon (1925) pointed this out, and it has been well established since (Steven & Carlisle 1959; Bunce & Jeffers 1977; Watson 1977; Nethersole-Thompson & Watson 1981). The exception at Glen Tanar followed a big reduction in deer numbers (Nethersole-Thompson & Watson 1981). The same general lack of appreciable natural regeneration applies to birch woods in upper Deeside, and is attributed to browsing by high densities of sheep, cattle and red deer (Brown & Wightman 1988).

The fact that fencing has been followed by considerable regeneration of pine and/or birch at Glen Tanar, Loch Muick, Creag Ghiubhais, Balmoral, Invercauld, Mar, and Mar Lodge indicates that browsing had previously prevented regeneration. Burning is not sufficient to remove regenerating trees except where they are concentrated on high-altitude moors with small trees. Burning on low-altitude moors next to large pines has been followed in some cases by the densest pine regeneration seen in the survey area, and on low moors with many young birches by a rapid spread of birch by seed and from the root.

Nethersole-Thompson & Watson (1974, 1981) stated that pines were colonising many Deeside moors outside the range of red deer. This was amply confirmed during the current survey, but the remit for the current report could not allow for including detailed local data on deer densities. During the survey, considerable areas with virtually no regeneration were also found outside deer forests, close to farms with large stocks of sheep and cattle on heavily-browsed grouse moorland and other heavily-browsed open upland. Much larger areas with virtually no regeneration lay on well-burned grouse moors, particularly at higher altitudes towards the potential upper woodland line.

Conservation advantages and disadvantages

If natural woodland in areas of conservation value were to extend on to adjacent open upland, this would make the future or such woodland more secure. The spread of regenerating woodland to form larger woodland blocks would also be expected, from the species-area relationship, to lead to more species of woodland flora and fauna per given area. A major extension of boreal woodland would probably lead to a spread of boreal flora and fauna. This may already be happening with wrynecks (Jynx torquilla) in Deeside, and other Scandinavian boreal forest birds would be expected to summer or breed there in future if the naturally regenerating woodland spreads.

If certain moorland areas are of high value for nature conservation, there is a case for removing natural tree regeneration there. The example of this at the Muir of Dinnet National Nature Reserve, in the survey area, has already been mentioned. Some nature conservationists consider that British moorland is a much more uncommon habitat type internationally than natural boreal forest, even though most of the moorland is man-induced and hence unnatural. Their grounds for this are that the moorland plant communities and bird communities are uncommon internationally, whereas natural boreal woodland is more common abroad, even if not in Britain. On the other hand, the nature conservation values of naturally regenerated forest exceed those of intensively afforested woodland (NCC 1986).

Extension of natural woodland on to rnoorland would decrease the potential for shooting of red grouse and mountain hares, but would increase it for shooting black grouse, and subsequently capercaillie and roe deer.

Natural regeneration and forestry

Changes in the practices of forestry have been advocated by several organizations including NCC, CCS, COSLA, Friends of the Earth and RSPB (e.g. Bain 1987). The present survey is a factual statement that may help rational discussion of these matters. In other countries with substantial natural regeneration, intensive planting is less important than in Britain and Ireland, where most woodland has had to be established on completely treeless ground. It is arguable that parts of Deeside and Donside with natural tree regeneration or potential for it could have different forestry practices from treeless parts of these islands. This may also apply to other parts of Scotland where natural tree regeneration is spreading on to open upland, notably Speyside, parts of Banffshire, moors in Morayshire, Nairnshire and eastern Inverness-shire, and some smaller moorland areas in easter Ross and eastern Sutherland.

Acknowledgments

I thank Manuel Hinge for putting data in a computer and doing analyses. Philip Rothwell, Clare Sampson, William Williams and David Bale of the Nature Conservancy Council for help with data abstraction and funding for Manuel

Self-sown Scots pines burnt, ploughing for planting spruce, Kerloch, April 1984

Hinge, and Robin Callander for notes on some estate boundaries.

References

Bain, C. (1987). Native pinewoods in Scotland. A review 1957–1987. Royal Society for the Protection of Birds, Edinburgh.

Brown, I.R. & Wightman, A.D. (1988). The birch woodlands of Deeside, 1974–1985 - a declining resource? Scottish Forestry 42, 93–103.

Bunce, R.G.H. & Jeffers, J.N.R. (1977). Native pinewoods of Scotland. Institute of Terrestrial Ecology, Cambridge.

Callander, R. (1987). A pattern of landownership in Scotland. Haughend Publications, Finzean, Aberdeenshire.

Gordon, 5. (1925). The Cairngorm Hills of Scotland. Cassell, London.

Kommission (1979). Das Schweizerische Tourismus-konzept. Grundlagen für die Tourismuspolitik Schlussbericht. Beratende Kommission für Fremdenverkehr des Bundesrates. Bern.

Krebs, C.J. (1978). Ecology: the experimental analysis of distribution and abundance, pp. 434–439. Harper & Row, New York.

NCC (1986). Afforestation and nature conservation in Britain. Nature Conservancy Council, Peterborough.

Nethersole-Thomson, D. & Watson, A. (1974). The Cairngorms: their natural history and scenery. Collins, London. Second edition (1981). Melven Press, Perth.

Steven, H.M. & Carlisle, A. (1959). The native pinewoods of Scotland. Oliver & Boyd, Edinburgh & London.

Tansley, A.G. (1953). The British Islands and their vegetation. Cambridge University Press, Cambridge.

Watson, A. (1977). Wildlife potential in the Cairngorms region. Scottish Birds 9, 245–262.

Watson, A. (1984). A survey of vehicular hill tracks in north-east Scotland for land use planning. Journal of Environmental Management 18, 345–353.

Watson, A. & Hinge, M. (1989). Natural tree regeneration on open upland in Deeside and Donside. 28 pp plus maps. Institute of Terrestrial Ecology, Banchory.

Table 1. Tree species found in the regeneration survey and their relative abundance (in terms of the number of kilometre grid squares containing regeneration of each species) on open upland in Deeside and Donside

		Deeside	Donside
Scots pine	Pinus sylvestris	484	362
Birch	Betula spp.	314	214
Rowan	Sorbus aucuparia	228	189
Willow	Salix spp.	149	84
Larch	Larix decidua	137	229
Norway spruce	Picea abies	41	76
Beech	Fagus sylvatica	13	33
Holly	Ilex aquifolium	16	0
Alder	Alnus glutinosa	11	2
Oak	Quercus spp.	6	6
Aspen	Populus tremula	7	1
Rose	Rosa spp.	5	0
Ash	Fraxinus excelsior	1	3
Hawthorn	Crataegus monogyna	2	0
Yew	Taxus baccata	0	1
Number of km squares with regeneration in		584	440

Table 2. Factors used to calculate the area occupied by each tree species according to its position in an order and the number of possible positions, in cases where two or more species were regenerating on an area of known size

	Position				
No. of	1	2	3	4	5
species					
2	0.666	0.333			
3	0.50	0.333	0.166		
4	0.40	0.30	0.20	0.10	
5	0.333	0.266	0.20	0.133	0.066
6	0.285	0.238	0.19	0.142	0.095

Table 3. Total (ha) of areas where regeneration comprised only one tree species, on open upland in Deeside and Donside (T = trace)

	Deeside	Donside
Pine	7717	3431
Birch	1775	246
Rowan	206	468
Larch	99	233
Willow	63	41
Holly	61	0
Norway spruce	16	20
Beech	7	19
Alder	3	T
Oak	2	6
Aspen	2	0
Rose	T	0
Ash	0	2
Hawthorn	2	0
Yew	0	1

Postscript. Changes since 1985

Since 1985, scattered trees and small patches of tree regeneration on some grouse moorland have become consolidated into dense mature woodland. Good examples are on Kerloch, Lamahip at Glen Cat, the Guard on the Fungle, parts of Glen Tanar, Mortlich, Muir of Dinnet, Crannach and Micras.

Burning, ploughing and dense tree planting have caused losses of regenerating trees on the west part of Kerloch, Stot Hill, Leadlich and Craiglich and Tirebagger, the west end of Bennachie next to the Brindy and the south-west part above Glenton, and part of the hill above Cattie on Whitehouse estate, the south side of Pressendye on Douneside estate, and slopes above Wisdomhow and Burn o' Vat at Muir of Dinnet.

There have been no natural material losses of regenerating trees, with one notable exception, on some grouse moors. There, estate staff have nearly eliminated regenerating young trees and scrub of juniper and other scrub species over large areas, by burning, pulling out seedlings, and felling saplings and older trees. The most conspicuous examples are on Carn Ferg, at Dinnet estate on the east slope of Culblean down to the road to Logie Coldstone as far north as Redburn, much of Geallaig Hill including slopes above Balanreich and a large area on Invercauld estate stretching from above Ballachrosk in Glen Gairn round via the East and West Milton Burn to the Strone, Crathie Burn and back east to above the Lebhall, including burning of the pine scrub above the natural timberline at Creag na Creiche, and a large area on Dinnet estate at Mammie, Glen Fenzie, the glen above Lary, hills east of the road above Lary. and across south-east to above Balmenach and Tullich Burn.

Several estates that practised destruction of naturally regenerating trees applied and received Forestry Commission grants of taxpayers' money to plant trees on other parts of their estates. This took place at Dinnet and Invercauld. Candacraig applied to plant a big area at Ladylea, eventually refused by the Forestry Commission because there would have been a risk to capercaillie from the new fences.

In Strath Don, regenerating pine was cut without a felling licence on Candacraig estate in the early 2000s, breaking Forestry Commission regulations. Stephen Brown of the FC witnessed the felling and spoke to the contractor. He made a report to the procurator fiscal, but the latter took no action, a biased inappropriate decision.

An area of 70.25 ha of regenerating pines on Dunecht estate was destroyed on the Hill of Fare in April 1997 by a muirburn fire that ran out of control.

Chapter 3. A critique of 'Norway and Scotland: a study in land use', a report based on the Reforesting Scotland Norway Study Tour May 1993

Introduction

'Reforesting Scotland' is 'a non-government organisation working to restore the land and communities of Scotland through reforestation' and 'a charity dedicated to ecological restoration and sustainable development in a well-forested land' (statements on the cover of its magazine The Tree Planters Guide to the Galaxy, later changed to Reforesting Scotland). It organised a tour to Hordaland in western Norway in May 1993 (Reforesting Scotland 1994). The tour was 'born of a belief that there is something to be learnt from a country such as Norway, with its ecological and climatic similarities to Scotland, which could be of benefit to Scotland'. It is useful for countries to learn from one another, but the Report did not define 'ecological and climatic similarity', and did not mention the material ecological and climatic differences between Scotland and south-west Norway. Below I discuss these issues after making a critical analysis of the Report, including pointing out certain biases and incorrect statements. I would like to see more woodland in Scotland, particularly near settlements and with a far greater emphasis than hitherto on naturally regenerated woodland. However, it is important that any ecological and climatic basis for this should be impartial and factual.

Critique of the Report

The Report's paired photographs depicting contrasts between Scotland and Norway are not properly comparable pairs, because the Scottish ones were taken during winter and mainly in dull conditions, and the Norwegian ones during summer in bright sunshine. They are therefore misleading in portraying Scotland unfavourably relative to Hordaland. The following comments apply:-

p. 5. The Scottish photograph is in winter, with leafless broad-leaved trees, dead brown rushes in the field, and brown moorland vegetation in winter condition on the rock-dominated ridges beyond the village. The Norwegian photograph is in early summer, with broad-leaved trees in full leaf, relatively short grass indicating young spring growth, and flowers in grass meadows.

p. 11. The Scottish photograph shows a summer scene, as manifested by green moorland vegetation, but the photograph was taken on a dull day and was so under exposed that everything bar the sky is too dark, producing an unrealistically gloomy effect. The Norwegian photograph shows summer sunshine and blue sky, with large potato plants growing on rich topsoil, probably near sea level to judge from the other photographs and maps of the area covered by the tour. The Scottish photograph shows an inland area on high moorland lying over glacial till that is too infertile and boulder-strewn to develop for agriculture, and is devoid of such topsoil.

pp. 12–13. The Norwegian photograph is in summer, with broad-leaved trees in full leaf and potato plants, and the Scottish photograph in winter, obvious from the leafless trees and dead rushes. The Norwegian photograph shows almost flat alluvial soils that have been cultivated, whereas the Scottish one shows cultivated soils at a steep gradient difficult for cultivation, with infertile soils developed on glacial till and with no alluvium. The Norwegian photograph shows a rock-dominated landscape except for the alluvium, the Scottish one less rock-dominated (only on the upper half of the photograph).

The caption for the Scottish photograph states that the croft is no longer worked, the croft house has been sold as a holiday home, 'dykes, single trees and field patterns are disappearing and the small area of fertile ground is degrading due to lack of management'. This implies that these features are disappearing since the croft stopped being worked and since the house became a holiday home, but no evidence is given on this. Indeed, the complete cover of grass on the line of the removed dykes would indicate dyke removal years before the photograph was taken. The photograph shows no evidence of single trees disappearing, or that 'fertile ground is degrading due to lack of management'. A scar of bare soil caused by sheep digging is evident on the steep bank above the sea, but it occupies a minute proportion of the area and is on ground too steep to have been worked by plough. Some rushes indicate pasture deterioration, but this affects only a small proportion of the fertile ground. In wet conditions in north-west Scotland, where this photograph was taken, past management by reseeding has frequently resulted in poaching of ground by stock animals and consequent invasion by rushes, i.e. a consequence of management rather than a consequence of lack of management. The term 'degraded' for the fertile ground is undefined and vague. It could imply degraded soil or degraded vegetation. There is no evidence for the former and doubtful evidence (as stated above) for the latter.

p.14. The Scottish photograph was taken in dull conditions outside the growing season, as is clear from the dead grass and the lack of fresh young growth in the coniferous trees, whereas the Norwegian one is in bright summer sunshine, with broad-leaved trees and grass in full leaf. The Scottish one shows dark green conifers behind, and the Norwegian one birchwood, but it would have been easy to show a Norwegian photograph with conifers and a Scottish one with birches.

p. 17. The Norwegian photograph shows a cloudless summer day, and the Scottish one a winter day (obvious from the low shadows and the dead moorland plants), and moreover was taken under a partly cloudy sky, resulting in large dark shaded areas on the hill.

pp. 22–23. The Norwegian photograph portrays a cloudless summer day, and the Scottish one a winter day (obvious from the dead moorland vegetation) and a sunless one, on a small piece of moorland near Lochstack Lodge in west Sutherland. The caption for the Scottish one incorrectly states 'Deserted farm'. In fact the building was a former hut for anglers fishing on nearby Loch Stack, and a footpath from the building leads to a pier at the loch-side close nearby, beside the mouth of the stream Allt Beul nan Cadhachan. The photograph shows no extensive topsoil with associated smooth grassland, such as would be found if the place had been a farm, but instead portrays moorland developed on infertile till above acidic ice-smoothed gneiss. The tiny strip of smooth grass round the building is typical, resulting from dung and urine from sheltering sheep along with urine and food scraps from anglers.

p. 9. This photograph in Norway shows 'sheep in winter on the coastal heath: a harsh oceanic climate'. There is thin snow cover, with the vegetation projecting well through it. The 'harsh' is misleading. Sheep can be kept outdoors there because the oceanic climate is mild with relatively little snow, and hence not harsh. The Report omits to explain that the coastal heath is kept treeless by sheep browsing and muirburn as in Scotland, yet states (p. 6) that the group heard a lecture by Prof Peter Kaland 'whose special interest is the ecology and history of the coastal heathlands which have been managed for 4000 years by farmers'.

p. 21. This single Scottish photograph shows a 'degraded cultural landscape' by Loch Tay. The term 'degraded' is not defined or obvious. The rest of the caption reads 'Two hundred years of intensive grazing by sheep have reduced a rich upland summer pasture to an impoverished mat-grass sward'. This sentence is misleading and largely speculative. To judge from events in other parts of the central and east Highlands, it is probable that many sheep have grazed this area for far more than 200 years. It is speculation to state what the pasture was like 200 years ago. It is overstatement that it is now 'an impoverished mat-grass sward', for the photograph shows high-quality smooth grass on the enclosed pasture, where sheep are at high density, and beyond it a complex mosaic of hill vegetation with very few sheep, much heath on freely drained parts, and mire vegetation on flushes below obvious spring-lines. The use of the term 'impoverished' is questionable, for it is not clear what is meant, e.g. whether the mat grass itself is impoverished in some way, or whether the sward has become poorer in palatability because of an assumed increase of mat grass. The brown colour of the bracken shows that the photograph was taken in winter, as also suggested by the heather being dark brown, the rushy mires pale, and the top of the hill having a cover of fresh snow.

Climatic and ecological differences, Scotland and Hordaland

Inspection of data held by the Norsk Meteorologisk Institut shows that the mean air temperature in July 1961–90 for Ullensvang, the fruit-farming district visited on the study tour, was 15C. Ulvik in the upper part of the fjord, further up than the easternmost ferry crossing on the tour, also had the same July mean, and Hjeltnes further up had 15.1C. Voss, also well away from the outer coast and to the north-west had 14.9, almost as high. This is markedly warmer than Scotland, where the Monthly Weather Report of the

Meteorological Office shows no station with such a high mean in 1961–90, the warmest value being 14.6 at Glasgow. The high July values at Ullensvang, Ulvik and Hjeltnes show that the southern part of inner Hordaland, where the tour members visited fruit farming, receives July warmth like central and south-east England, not Scotland.

The 15C July isotherm includes a wide strip along the southern Norwegian coast as far north as Haugesund, whereas in Britain it runs through east Yorkshire and south Cumberland, and then from north of Dublin southwards to include the south Irish coast. Francis's (1981) review of climate in the main farmland areas of Scotland provides useful information on mean temperature, length of growing season, and other factors relevant to the growth of crops. Only two agricultural areas, which are both in the most sheltered part of central Scotland, had a July mean above 14.5, in each case 14.8.

In Norway, even Trondheim far to the north of Hordaland has a July mean of 14, and Bergen in Hordaland 14.4 (Hansen 1961). Hansen stated that Norway has 4–5 months of winter (mean monthly temperature <0C), except for the south-west coastal strip, which has no winter as thus defined. The January mean 0C isotherm runs through Hardanger Fjord, the top of the fjord being colder but the outer part of it and the coastal islands having no winter. All of Scotland at sea level has a January mean temperature above the 0C isotherm (and hence no winter as defined above), and the 5C isotherm runs through Mull, Kintyre, and North Uist, with milder conditions next to the Atlantic.

Norway is less windy than Scotland, except on the Norwegian outer islands and coast, but they in turn are less windy than the outer Scottish islands and coasts. The inner parts of Hardanger Fjord such as Ullensvang are very sheltered from the prevailing winds off the North Sea, by reason of distance from the coast and intervening mountains towards the coast.

Ullensvang and neighbouring areas lie in the rain shadow of the large Folgefonn icecap (which reaches 1660 m) and many mountains above 1400 m, higher than any in Scotland. This leads to dry Fohn or Chinook conditions east of the mountains, aiding in the drying of soil moisture and hence favouring early cultivation and growth of crop plants.

Greater oceanicity in Scotland than in Norway is evident also in the distribution of wild plants and animals. The northern limit of Scots pine (Pinus sylvestris) in Scotland is in the southern part of Sutherland, whereas in Norway it extends north to Porsanger Fjord in Finnmark, well to the north of 70 degrees N and far beyond the Arctic Circle. In Scotland the natural range of the beech (Fagus sylaticus) did not extend further north than Edinburgh at 56N, but in Norway it went to about 61N, almost to Sogne Fjord. The garden warbler (Sylvia borin), uncommon in Scotland north of Perthshire and absent from much of northern Scotland including the Highlands, abounds in suitable habitat as far up as northern Troms county and Finnmark county, well to the north of 70N and far beyond the Arctic Circle.

The relevance of climatic differences to the Report

The Report does not mention that farmers in an oceanic climate can keep far more animals per ha on deforested moorland that is managed as pasture, than is possible in woodland (see Appendix below). In more continental conditions with snow, such as in Scotland in upper Strath Don or in interior parts of Hardanger Fjord, stock cannot be wintered out of doors unless they are provided with supplementary food that has been grown as a crop. In such cases, stock numbers are limited by the amount of winter keep that can be grown on fields, which in turn is limited by the amount of suitable land locally. If such feed is imported from outside, it must be bought, and the economics of purchase, transport, storage and handling of feed are marginal.

Slimon (1993), the wife of a hill farmer in the central Scottish Highlands and a member of the Tour party, described an inland farm in Hordaland that had 4.1 ha of agricultural land and 59.9 ha woodland, as well as much mountain land. Noting that 'The sheep and cows were in-wintered and only going out at the end of May, early June', she added about her farm in Scotland that 'we could grow more trees but not at the expense of our cattle and sheep'. This last remark was consistent with the fact that one can keep more stock per ha on deforested moorland than in woodland. Gordon-Duff (1993), also a member of the Tour party, stated that the average farm in the area of the Norwegian study tour had only 10 ha of agricultural land and 56 ha of woodland, so the one described by Slimon was typical in being small compared with Scottish hill farms.

Because of shading, understory vegetation under pinewood (and even under birchwood except in very open woods) has a far smaller biomass than on treeless moorland and pasture, and much of the understory grass is of lower nutritive value. A high proportion of the vegetation, especially under pinewood, comprises moss and lichen, which sheep and cattle avoid eating.

The climatic differences mentioned above are evident in different parts of Scotland and Norway. In the less oceanic parts of Norway, such as Hardanger Fjord as compared with the coastal islands, winters are colder with more snow, in-wintered stock numbers are limited by the small amounts of arable land available to grow winter keep, and resulting summering stock numbers are so low that there is under-grazing of summer pasture and insufficient browsing to prevent natural tree regeneration. Scottish and Norwegian meteorological data demonstrate the climatic differences, with inland sites such as Aboyne and Ullensvang showing lower winter temperatures and more snow than at the outer coast next to the North Sea.

Geomorphological differences

The Report states (p. 9) about Norway that, 'Like Scotland, it is a land dominated by mountains, forests and the sea'. This is inaccurate and misleading. It omits the fact that Scotland has large tracts of productive arable land with very little woodland, a land type absent from Norway except for a relatively small area near Stavanger. This fundamental difference occurred because Norway had colder winters and was more mountainous and also with higher mountains, which resulted in a more heavily glaciated rock-dominated landscape than is typical in most parts of Scotland other than the western Highland mainland and Hebrides.

An obvious point from the Report's photographs, and from inspection of ground in Norway and Scotland, is that the Hordaland glaciers shifted most of the till, morainic material, and fluvio-glacial material into the sea, as in similarly heavily glaciated parts of the west Scottish Highlands such as the Lorne area of Argyll. The Norwegian photograph on p. 12 shows an ice-smoothed rocky landscape with no substantial glacial deposits, the small amount of farmland being on flat alluvial topsoil. The photograph of a Norwegian potato plot on p. 11 again shows flat topsoil, though the photographs on pp. 5, 14, 17, and 23 do show glacial material, albeit very localised and small on p.11, and not fairly extensive except on p. 5. The Scottish photographs on pp. 5 and 13 show fairly extensive farmland developed on till in Sutherland. Nearly all of the Norwegian photographs, and especially the frontispiece and pp. 5, 9, 12 and 14, show a rock-dominated landscape with very little land that would be suitable for pasture or arable crops.

Economic differences

The Report draws attention (p. 11) to the higher population density and bigger main town in Hordaland than in Highland Region, despite Hordaland having far less cultivated farmland and productive woodland. This ignores the relative importance of sea fishing and industrial development, including Norway's long global dominance in mercantile shipping. To imply, as the Report does, that the Highlands could have a higher population density and bigger main town than Hordaland because the Highlands have larger areas of cultivated farmland and productive woodland is misleading.

When describing 'financial support' (a euphemism for taxpayers' subsidies and grants) for Norwegian agriculture, the Report omits to mention that Norwegian farmers were at that time among the most highly subsidised per head in the world, but does state correctly (p. 15) that Norway provides a 'Producer Subsidy Equivalent (PSE) of 77% compared with 47% for the EC'. At the time of the Report, there was already increasing opposition to these high subsidies by urban Norwegian taxpayers who supply the country's main wealth. The Report states that 80% of the Common Agricultural Policy funds that reach farmers in the UK go to the largest 20% of farms, but omits to mention the proportions per ha (e.g. if the largest 20% of farms occupy 80% or more of the total area of fields, this would put a different light on the distribution of subsidies).

On p. 15, one of the Report's main conclusions is that the 'Norwegians do have one great advantage over us in Scotland – they have a definable, manageable, politically supported and understandable land use system which although it has evolved over many years has a clear logic to it'. It is questionable whether the urban Norwegian taxpayer would have agreed that this is a great advantage. Given the damaging effects of the CAP-driven export subsidies on third-world agriculture as well as on pollution of groundwater and food, and adverse impacts on landscape and wildlife and bees, there are already trends to reducing or eliminating production subsidies and export subsidies. Hence it is increasingly likely that all EU agricultural subsidies will eventually end. It therefore seems surprising that the Report is so out of date in praising high subsidies. In this respect the Report uses the term 'clear logic', but logic, whether clear or otherwise, does not arise in this case. The authors' use of the term is erroneous and misleading.

A news item on Grampian Television in northern Scotland on 18 April 1998 showed a group of Norwegians from Hordaland on a tour of Caithness, the aim being to get ideas how to stem rural depopulation, which has become a serious problem in Hordaland. The high rate of emigration of educated young people who have finished schooling was emphasised. This is a world-wide problem in remote rural areas with marginal agriculture, especially in mountainous land (International Mountain Conference 1996). The fact that it is severe in Hordaland, despite very high subsidies, indicates the difficulty. Spokesmen said that there was no interest in joining the EU, but did not discuss whether this might be due to the EU's lower subsidy rates.

Acknowledgements

I thank Dr Audun Stien for his interest and the Norsk Meteorologisk Institut for sending data.

References

Caulfield, S. (1981). Forest clearance and land use in Mayo around 3,000 B.C. Irish Forestry 38, 92–100.

Francis, P.E. (1981). The Climate of the Agricultural Areas of Scotland. Climatological Memorandum 108, Meteorological Office, Edinburgh.

Hansen, S.W. (1961). The climate. Norway North of 65 (ed. by O. Vorren), pp. 37–49. Oslo University Press.

International Mountain Conference (1996). Proceedings of conference at Aviemore, Scotland, spring 1996.

Kaland, P.E. (1986). The origin and management of Norwegian coastal heaths as reflected by pollen analysis. Anthropogenic Indicators in Pollen Diagrams (ed. by E.E. Nehre), pp. 19–36. Blakema, Rotterdam.

Reforesting Scotland (1994). Norway and Scotland: a study in land use, a report based on the experience of the Reforesting Scotland Norway Study Tour May 1993. Reforesting Scotland, PO Box 1707, Edinburgh (ISBN 0 9523012 0 2).

Slimon, S. (1993). Norwegian study tour 1993. Rural Forum 31, 4.

Watson, A. (1989). Land use, reduction of heather, and natural tree regeneration on open upland. Report of the Institute of Terrestrial Ecology for 1988-89, 25–27.

Watson, A., Phillips, J. & Moss, R. (1995). Moorland history and grouse declines in Ireland, with an outline policy for conservation and recovery. Canada Life Irish Red Grouse Conference (ed. by R. O'Dwyer), pp. 26–100. Irish Field Trials Grouse Ground Committee, Cappagh, Waterford.

Appendix from p. 36 of Watson et al. (1995).

Farmers cleared most woodland in Ireland and Britain, but not in Norway or Finland. Gentler land relief, higher soil fertility and the more oceanic climate in Ireland and Britain favoured the spread of fields. This explains why only 3% and 10% of Norway and Finland support agriculture, but 68% and 77% of Ireland and in Britain (the British value including 22% in the intermediate Scotland). Forest clearance for farm stock supported more people (estimated at 300-fold) than hunting in forest (Caulfield 1981). Those who favour large-scale tree planting compare Ireland's 7% woodland cover with Norway's 28%. Clearly, the comparison is flawed.

On land unsuitable for fields, the key difference is climatic. In the mild winters of Ireland and Britain including Scotland, hill stock can be wintered outside, feeding largely on moorland vegetation. Palatable grass in fields and on the moor grows even in winter during mild spells, and other moorland plants provide evergreen feed.

In Finland and all of Norway except the most oceanic coasts, winters are too severe for stock to winter outside. Flock size is limited by the amount and quality of winter feed that can be grown in fields during the short summer. In turn, this is limited by a shortage of fertile soil for arable and grass cropping, often associated with surface bedrock and steep slopes unsuitable for cultivation.

However, in the mild winters of oceanic parts of western Norway, much land has been deforested and there is a long tradition of moor burning (Kaland 1986) and cutting trees and heath plants for winter fodder. Sheep are wintered outdoors. In recent years a decline of farming and consequently fewer grazing animals have led to some reversion to birchwood. This, along with reversion to naturally regenerated pinewood, especially on the more infertile soils, has occurred also in the more continental parts of north-east Scotland such as Deeside and Donside, in association with a similar decline of hill farming and concomitant reduction of stock (Watson 1989).

Table 1. Mean air temperature in July at two settlements visited in the Norwegian Tour and at three Scottish settlements, all five at sea level.

	July mean temperature (C)
Ullensvang	15.0
Hjeltnes	15.1
Aberdeen	13.7
Plockton (Wester Ross)	13.8
Inverness	14.5

South side of ancient Scots pine, Glen Quoich, June 2012

North side of same pine

Chapter 4. Highland sporting estates and land 'impoverishment'

At times one reads that Highland sporting estates impoverish land. There are several social and economic reasons for having no private estates in Scotland, as in Norway or Switzerland, but land impoverishment is not one of them. Dr Frank Darling's historical account of Scottish deforestation was overstated, but has been the basis for much invalid comment since. Recent examples are by Sir John Lister-Kaye in Scottish Wildlife and Sir William Gordon Cumming in Scottish Forestry, claiming unsustainable impoverishment of moorland on sporting estates. Their statements were anecdotal allegations without supporting evidence or scientific foundation. I wrote to both journals, briefly stating the scientific evidence that rejects these allegations. Such is the bias of these two journals that neither letter was printed.

Most Scottish deforestation occurred so long ago that we lack written record of it. Maps in the 1600s show a woodland area similar to that of the 1800s when sporting estates began, and that of today if we omit plantations. Deforestation was caused mostly by the climate becoming more oceanic and at low altitudes largely by clearance for farm stock.

Muirburn on sporting estates maintains deforested land in a treeless condition, but less so than on the same land in earlier centuries when farmers did the burning before sport shooting became important, and less so than on today's private sheep land or crofters' communal grazing land. On grouse moors, muirburn generally has impacts that are less adverse to wildlife and soil than the generally low standards of muirburn on communal croft grazing land. Where communal rough grazing land overlies peaty soils above hard acidic bedrocks such as quartzite, in areas with heavy rainfall, poor muirburn combined with high stocking rates on land with much bared peat due to severe grazing at high stocking rates can result in rapid runoff and severe peat erosion.

I saw this alarming process taking place on communal grazing land at Milltown near Glenamoy in northern Co Mayo in April 1967, along with P.J. O' Hare, then Director of the Peatland Experimental Station of An Foras Taluntais (The Agricultural Institute) at Glenamoy. Again I watched it occurring in September 1996 on the slope west of Laid at Loch Eriboll in north-west Sutherland, along with soil surveyor Sandy Walker, ecologists Stuart Rae and Des Thompson, and John Phillips the then Director of the Heather Trust. In both cases the entire hillsides became awash with small streams and rills, flowing quickly downhill and visibly stripping off pieces of bared peat up to 3 cm in length. The streams lower down the slope were dark in colour because of their high content of peat in small lumps as well as innumerable suspended particles.

In the early 1980s I was informed by expert observer Roger O'Donovan, then a senior officer at Glenmore Lodge, who witnessed the same events on the hills of Harris in the Outer Isles. I have not seen it on sporting estates, except in Glen Feshie where high stocking density of wild red deer was the problem, without muirburn.

South side of old Scots pine, Loch Davan, Jan 2012

North side of same pine, January 2012

Lichen on Scots pine roots, moister near soil surface, Loch Davan, March 2012

Other cases observed by me were on two Kincardineshire deer farms at Glensaugh and Fasque, again with high stocking and much bared ground, not due to muirburn.

Farmers' sheep graze all grouse moors save those with tick-borne disease. The worst moorland de-vegetation, soil erosion, and deposition of eroded matter in rivers to be seen in these islands has occurred since 1970 on communal sheep grazing land in Co Mayo, western Ireland. It followed a large increase of hill sheep following Ireland's entry to the EEC, in association with high headage subsidies per ewe, as pursued by the Common Agricultural Policy. It occurred on land where there have been no sporting estates for decades.

Chapter 5. Orientation by plants and animals

Trees

Many decades ago I noticed that the bark on the north side of the lower bole or trunk of any Scots pine (Pinus sylvestris) tree in northern parts of Norway, Sweden and Finland, and repeatedly in upper Deeside of north-east Scotland, carries much lichen, whereas the bark on the south side of the same tree is devoid of lichen. The colour difference is striking, brownish-red clean bark on the south side, greyish-green lichens sprouting on the north side a few inches away. The bark on the upper half of the bole tends to be red and mainly lichen-free, irrespective of the orientation. I found the same features later in other tree species as described below, and in the boreal forest of Quebec, Newfoundland and Minnesota. This offers a useful sign for anyone who wishes to know directions in a wood, on occasions when dark clouds obscure the sun and the stars at night, and the person lacks a compass and map, and also a watch.

I have looked at several books on northern survival and wilderness survival, covering North America and northern Europe and northern Asia, as well as covering places at high altitude (e.g. Berglund 1972; Department of Indian Affairs 1972). To find direction, the authors stressed compass and map. They also wrote that if the sun is shining one can use a watch to check against the highest point attained by the sun, at around noon, or can poke a stick into the ground and measure the length of shadow

North (left) and south (right) sides of old larch Birkhall, August 2012

Lichen on lower parts of Scots pines, no lichen on red bark higher, Blelack, October 2011 (Derek Pyper)

South side of oak Craigendarroch, 15 June 2012

North side of same oak

South and north sides of same oak near ground

cast by the stick, the shortest length being at noon. Also mentioned was the sighting of the Pole Star. None of them mentioned lichen or moss on trees.

The difference described above for Scots pine holds widely in north-east Scotland also for old European larch (Larix decidua), for instance at Loch Davan and Allt Cailleach woods near Birkhall. It applies also to old Norway spruces (Picea abies), such as those beside the gate into the Forestry Commission's wood at Brathens near Banchory, beside the junction of the roads to Glassel and Hill of Brathens, and at Glen Tanar. I found that it applies also to old oaks (Quercus robur), such as near Cairnton at Crathes and in the old oak wood (Site of Special Scientific Interest) at Craigendarroch above Ballater, Likewise I found it conspicuous on silver firs (Abies alba) at Glen Tanar.

In dense young Scots pines within a thicket it is less conspicuous or even absent, with lichen growing all round the lower bole of the tree. Later, as some trees die because of competition from shading, the boles are more in the open and the lichen begins to predominate on the south side. Not every old pine shows the big difference between the north and south side, because there are local variations such as proximity to an opening such as a track or public road or marsh or river or loch.

Orientation involves many

South side of old Norway spruce, Brathens, June 2012

North side of same Norway spruce

South side of same Norway spruce, near ground

common lichen species that grow on trees, such as Evernia prunastri, Hypogymnia physodes, Ramalina farinacea. The longer trailing Usnea species (often called beard lichens) and Bryoria species are more frequent in upper Deeside than at Ballogie in mid Deeside and Glen Dye in lower Deeside, and tend to be concentrated on the north sides of the trees.

Silver birch (Betula pendula) in lower, mid and upper Deeside has lichens growing abundantly on all sides of the main lower bole of the tree. Occasionally I have noticed a difference between the north and south sides, with more lichen on the north side, but slight compared with that on Scots pine, larch and oak. Nevertheless, on ancient silver birches, such as those beside the foot of the road to the car park in the Forestry Commission's woodland at Cambus o' May, it is strikingly obvious.

Reindeer lichens (several Cladonia species, especially Cladonia ciliata, and sometimes Cetraria islandica) growing on the ground are usually most prolific on the south side of a pine tree, except in extremely open spots, often though not always exposed to the prevailing south-west wind, where the ground lichens below the trees may face the exposed direction, and the lichens on the lower boles or trunks may be on the more sheltered side. If the opening is to the north, because a marsh or loch or road prevents trees growing in

South side of silver fir, Glen Tanar, 13 July 2012

North side of same silver fir

South side of same silver fir, near ground

that direction, then the lichen-free parts of the old pines or larches face the exposed direction.

In dense heavily-shaded woodland, such as a beech (Fagus sylvatica) wood or a plantation of Sitka spruce (Picea sitchensis), algae and mosses tend to abound more on the north sides of the boles or trunks of the trees than on the south sides. Wherever a tree is leaning off the vertical, or has branches spreading sideways, mosses thrive on the flattest parts or parts at low gradient.

A general feature in all the tree species mentioned above is that lichens also grow on the south side of the tree, indeed all round it if you look near the ground, presumably favoured by greater and more frequent moisture just above the ground. This applies to the first vertical foot off the ground, occasionally more. Also, moss sometimes grows on the south side, always next to the ground and within the first six vertical inches from it.

Alpine plants

It has long been well known that alpine plants tend to be prostrate, creeping low along the ground. This is a reaction to strong frequent winds. As one moves from a sheltered spot to an exposed ridge, the proportions of prostrate vegetation and bare ground increase. Inspection of individual plants reveals that the front of the plant grows in the opposite direction to

South side of old birch, Cambus o' May, 11 September 2012

North side of same tree

South side of second old birch, Cambus o' May, 11 September 2012

North side of second old birch

the prevailing wind, while the back of the plant is slowly dying. It is a dynamic process, each plant gradually moving away from the prevailing wind, as does the strip of bare grit between one plant and another.

Often the plants on alpine slopes, such as in the Cairngorms, are in the form of terracettes, with grit on the tiny terrace and plants along the vertical part of the step. This gives the appearance of waves, and again the terracettes are dynamic, moving in relation to the prevailing wind.

On the northern Cairngorms the prevailing wind is a katabatic wind coming from the high cold plateau to the south. Hence it is usually southerly, even though the prevailing wind in the main valley by the River Spey is from the south-west.

In June 1998 I took part in a site visit to wind turbines recently erected on exposed moorland hills at Novar Estate in Easter Ross. The development company had erected many white cloth tatter flags pinned to posts, to find which sites were windiest. The extent of tatter was measured over a period of a year or two. The company would then take decisions on locating sites for other turbines, using the evidence from the tatter flags. I pointed out to those present that the orientation of the growth of prostrate heather and other plants fitted the results of the tatter flags in some cases, but in others not, in the most extreme cases up to 90 degrees out. I explained that the plant orientation revealed the long-term (decades-long) response to wind on the site, hence a far better and less expensive method than frequent measurements of flags over a very short period of a year

Moss campion (Silene acaulis) showing 'compass effect' on a flat site, where flowers on the south side of a clump on a flat site open about two weeks earlier than on the north side, 70 degrees latitude, Gaaseland, East Greenland, 28 July 1994 (Fritz Schwarzenbach)

Polar willow (Salix arctica) gains heat from a south-facing rock in Owl Valley, Baffin Island, allowing part of the plant to produce earlier larger leaves above the rock, July 1953

Crowberry (Empetrum hermaphroditum) grows away from the prevailing wind on Cairn Gorm, its stem lying within a groove in bare grit, its leaves gaining heat from a rock, 20 September 1995

Adam Christopher Watson, 12 years old, at south-facing anthill, Creag Clunie, March 1975

or two. The turbines would be up for 25 years. Therefore, the use of flags for a year or two was an unreliable guide, because the time duration formed too small a proportion of the 25 years (only 4% or 8%). Also, one brief visit would be enough to measure plant orientation, whereas the checking of many flags requires many visits and more time per visit.

Arctic plants behave differently in other ways, according to the direction of the sun. An example of this can be seen in what the eminent Swiss botanist and Arctic explorer Fritz Schwarzenbach calls the 'compass-flowers'. The south half of a flower clump or tussock, such as moss campion (Silene acaulis) came into flower in the valleys of Baffin Island two weeks earlier than on the north side of the same tussock. He noticed similar effects in north-east Greenland. Despite daylight round the clock, the sun is higher in the sky when shining in the south, and creates more warmth through solar radiation reaching the plants and the soil around them. Another feature I noticed on Owl Valley in Baffin Island. A polar willow (Salix arctica) plant growing across the south-sloping side of a rock grew fresh catkins a week or two before other polar willows growing nearby but not across the surface of rocks. When I put my hand on the rock during a sunny day in late June, I could feel the remarkable warmth. Dark in colour because of dark lichens, some of which had an almost black hue, the rock gained extra heat from its dark surface by absorbing much incoming radiation.

Anthills

Soon after noticing the lichen-free south sides of pine trees in north-east Scotland, I spotted that anthills of the red wood ant (Formica rufa) under or beside pine trees or birch trees have a steep face on the north side and a low gradient on the south side. The ants construct the anthill in a way that maximises the surface exposed to warm sunlight in the middle parts of the day.

I showed this to Ian Murray of Ballater. Later, Ian attended at Glenmore Lodge Outdoor Centre in Strathspey a course that leads to a qualification in mountain leadership if candidates are successful. The staff are experts in navigation, and the qualification is highly regarded. During the course, navigation including orientation was covered carefully, involving compass and map. Ian happened to mention anthills. The staff member had not noticed the orientation of anthills and knew no one at the Lodge who had mentioned it. Surprised, he asked Ian, "Who told you that?" Ian replied "Adam Watson, and he showed it to me with anthills in woods on Deeside and Speyside." Ian was successful in the course, and duly gained his qualification in mountain leadership.

References

Berglund, B. (1972). Wilderness survival: a complete handbook and guide for survival in the North American wilds. Charles Scribner's Sons, New York.

Department of Indian Affairs and Northern Development (1972), prepared by the Employment and Related Services Division. Northern survival: a practical guide to survival in the northern regions. Information Canada, Ottawa.

Chapter 6. UK policies on Biodiversity and Sustainable Forestry

Biodiversity. The UK Action Plan (Command 2428, 1994), price £18.50, HMSO, London.

This book stated the UK Government's Policies on biodiversity and the Biodiversity Action Plan, and was signed by the UK Secretary of State for the Environment, and the Secretaries of State of Scotland, Wales, and Northern Ireland.

p. 100, paragraph 6.48. 'Action. Continue to protect ancient semi-natural woodlands and encourage forms of management which conserve their special characteristics. Encourage the extension and creation of native woodlands. Support the creation of community woodlands near population centres'.

p. 163, paragraph 10.15. 'Policies and programmes have as their objective to maintain the present area of ancient and semi-natural woodlands; to expand the area of native woodland of a natural character'.

Some other books by the author

1963. Mountain hares. Sunday Times Publications, London (by AW & R. Hewson)
1970. Animal populations in relation to their food resources (Editor). Blackwell Scientific Publications, Oxford and Edinburgh
1974. The Cairngorms, their natural history and scenery. Collins, London, and 1981 Melven Press, Perth (by D. Nethersole-Thompson & AW)
1975. The Cairngorms. Scottish Mountaineering Club District Guide, published by Scottish Mountaineering Trust. Second edition published 1992
1976. Grouse management. The Game Conservancy, Fordingbridge, and the Institute of Terrestrial Ecology, Huntingdon (by AW & G.R. Miller)
1982. Animal population dynamics. Chapman and Hall, London and New York (by R. Moss, AW & J. Ollason)
1982. The future of the Cairngorms. The North East Mountain Trust, Aberdeen (by K. Curry-Lindahl, AW & D. Watson)
1984. The place names of upper Deeside. Aberdeen University Press, Aberdeen (by AW & E. Allan)
1998. The Cairngorms of Scotland. Eagle Crag, Aberdeen (by S. Rae & AW)
2008. Grouse, the grouse species of Britain and Ireland. Collins, London, New Naturalist Library No 107 (by AW & R. Moss)
2010. Cool Britannia, snowier times in 1580–1930 than since. Paragon Publishing, Rothersthorpe (by AW & I. Cameron)
2011. It's a fine day for the hill. Paragon Publishing, Rothersthorpe
2011. A zoologist on Baffin Island, 1953. Paragon Publishing, Rothersthorpe
2011. Vehicle hill tracks in northern Scotland. The North East Mountain Trust, Aberdeen, published imprint Paragon Publishing, Rothersthorpe
2011. A snow book, northern Scotland: based on the author's field observations in 1938–2011. Paragon Publishing, Rothersthorpe
2012. Some days from a hill diary: Scotland, Iceland, Norway, 1943–50. Paragon Publishing, Rothersthorpe
2012. Human impacts on the northern Cairngorms: A. Watson's scientific evidence for the 1981 Lurcher's Gully Public Inquiry into proposed Cairn Gorm ski developments, and associated papers on people and wildlife. Paragon Publishing, Rothersthorpe
2012. Birds in north-east Scotland then and now: field observations mainly in the 1940s and comparison with recent records. Paragon Publishing, Rothersthorpe (by AW & Ian Francis)
2013. Place names in much of north-east Scotland. Hill, glen, lowland, coast, sea, folk. Paragon Publishing, Rothersthorpe
2013. Points, sets and man. Pointers and setters, stars of research on grouse, ptarmigan and other game. Paragon Publishing, Rothersthorpe

2013. Hill birds in north-east Highlands. Field observations over decades – ptarmigan, red grouse, golden plover, dotterel, bird counts. Paragon Publishing, Rothersthorpe

2013. Mammals in north-east Highlands – red deer, mountain hares, others. Paragon Publishing, Rothersthorpe

Author

Adam Watson was brought up at Turriff in north Aberdeenshire. In 1939 he developed a deep interest in the Cairngorms and surrounding area, and became a mountaineer and ski-mountaineer in boyhood. His frequent visits to that region led to the field observations that formed the basis of the papers in this book.

Lightning Source UK Ltd.
Milton Keynes UK
UKOW06n0603161014

240142UK00004B/22/P

9 781782 221890